THE COLD WAR CHRONICLES

NATO, the Warsaw Pact, and the Iron Curtain

Erik Richardson

New York

Published in 2018 by Cavendish Square Publishing, LLC
243 5th Avenue, Suite 136, New York, NY 10016

Copyright © 2018 by Cavendish Square Publishing, LLC

First Edition

No part of this publication may be reproduced, stored in a retrieval system, or transmitted in any form or by any means—electronic, mechanical, photocopying, recording, or otherwise—without the prior permission of the copyright owner. Request for permission should be addressed to Permissions, Cavendish Square Publishing, 243 5th Avenue, Suite 136, New York, NY 10016. Tel (877) 980-4450; fax (877) 980-4454.

Website: cavendishsq.com

This publication represents the opinions and views of the author based on his or her personal experience, knowledge, and research. The information in this book serves as a general guide only. The author and publisher have used their best efforts in preparing this book and disclaim liability rising directly or indirectly from the use and application of this book.

CPSIA Compliance Information: Batch #CS17CSQ

All websites were available and accurate when this book was sent to press.

Library of Congress Cataloging-in-Publication Data

Names: Richardson, Erik, author.
Title: NATO, the Warsaw Pact, and the Iron Curtain / Erik Richardson.
Description: New York : Cavendish Square Publishing, [2018] | Series: The Cold War Chronicles | Includes bibliographical references and index.
Identifiers: LCCN 2016049422 (print) | LCCN 2017000066 (ebook) | ISBN 9781502627278 (library bound) | ISBN 9781502627216 (E-book)
Subjects: LCSH: North Atlantic Treaty Organization--History--20th century--Juvenile literature. | Warsaw Treaty (1955)--Juvenile literature. | Communist countries--Relations--Western countries--Juvenile literature. | Western countries--Relations--Communist countries--Juvenile literature. | Cold War--Juvenile literature. | World politics--1945-1989--Juvenile literature.
Classification: LCC UA646.3 .R47 2018 (print) | LCC UA646.3 (ebook) | DDC 355/.03109182109045--dc23
LC record available at https://lccn.loc.gov/2016049422

Editorial Director: David McNamara
Editor: Jodyanne Benson
Copy Editor: Rebecca Rohan
Associate Art Director: Amy Greenan
Designer: Alan Sliwinski
Production Coordinator: Karol Szymczuk
Photo Research: J8 Media

The photographs in this book are used by permission and through the courtesy of: Cover Pudelek (Marcin Szala)/Wikimedia Commons/File:Czov (Zaisa) – preserved part of Iron curtain.jpg/CC BY-SA 3.0; p. 4, 61, 86 Bettmann/Getty Images; p. 8 Heritage Images/Hulton Archive/Getty Images; p. 10 TRAJAN 117/Wikimedia Commons/File:Austro-Hungarian Monarchy (1914).svg/CC BY-SA 3.0; p. 14 Imagno/Hulton Archive/Getty Images; p. 19 Print Collector/Hulton Archive/Getty Images; p. 21 Nellmac/iStock; p. 24 Peteri/Shutterstock.com; p. 26, 58, 63 Popperfoto/Getty Images; p. 28 52 Pickup/Wikimedia Commons/File:Map-Germany-1945.jpg/CC BY-SA 2.5; p. 34-35 Keystone-France/Gamma-Keystone/Getty Images; p. 37 Richard Nebesky/robertharding/Getty Images; p. 41, 47 Hulton Archive/Getty Images; p. 44 Spencer Platt/Getty Images; p. 50 Underwood Archives/Archive Photos/Getty Images; p. 65 ullstein bild/Getty Images; p. 66 OlegAlbinsky/iStock; p. 69 Gjon Mili/The LIFE Picture Collection/Getty Images; p. 70 Sovfoto/Universal Images Group/Getty Images; p. 72 Giorgio Galeotti/Wikimedia Commons/File:United Nations – New York, NY, USA – August 18, 2015 02.jpg/CC BY-SA 4.0; p. 75 Peter Hermes Furian/Shutterstock.com; p. 84 Hulton Deutsch/Corbis Historical/Getty Images; p. 89 Interim Archives/Archive Photos/Getty Images; p. 91 U.S. Air Force Photo/Wikimedia Commons/File: Sputnik 1.jpg/CC PD; p. 93 Ronald Reagan Library/Wikimedia Commons/File:Reagan & Gorbachev Arrive (8002548794).jpg/CC BY-SA 2.0; p. 96-97 Gerard Malie/AFP/Getty Images.

Printed in the United States of America

CONTENTS

Introduction:
What's Cool About a Cold War? 5

1 Mounting Tensions 11

2 NATO, the Warsaw Pact, and
  the Iron Curtain................. 27

3 The Faces of the Cold War 51

4 Negotiation and Innovation 73

5 The Legacy of NATO, the Warsaw Pact,
  and the Iron Curtain............. 87

Chronology..................... 102

Glossary 103

Bibliography................... 106

Further Information 108

Index 110

About the Author 112

A picture of Lenin watches over a public display promoting the glory of the Soviet military.

INTRODUCTION

What's Cool About a Cold War?

By the end of World War II, almost seventy percent of the industrial foundation of Europe had been destroyed. On top of this, millions of lives—soldiers and civilians—had been lost. Rather than sit back and rest and rebuild the economies and lives that had been shattered, though, the **Soviet** government's dreams of becoming an empire would spark new fears, and the wind-down of the military machine that normally comes in the wake of war would be postponed indefinitely.

The Soviets Wanted More

As the Cold War unfolded and began to find its stride, the dynamics became more complex. However, to give a fair sense of how it started, we can focus on three actions that reflect different levels of Soviet ambition at the end of World War II.

The Soviets seemed pretty fired up about the idea of expanding. In chapter one we see that in a slightly larger historical context. At the lowest level was the situation where the USSR (Union of Soviet Socialist Republics, as the Soviet Union was once called) made a determined effort to push tanks and troops into as many of the Eastern European countries as it could in order to exert power and influence over them.

At the second level was a concerted effort by the Soviets to abduct as many German military scientists as possible at the end of the war. Thousands of nuclear scientists, rocket scientists, and other researchers, along with their families and whole buildings of equipment, were transported out of Germany and taken back to the Soviet Union. Taken by itself, this might not seem like more than merely working to keep themselves in the race against countries like the United States and Britain in terms of weapons development. However, in light of the next action, this one deserves to be understood as more dangerous.

At the third level, Soviet leader Joseph Stalin proved he was willing to abandon any pretense at cooperation regarding conquered territories. He threw out important agreements, like those worked out at Yalta, when he did not allow democratic processes in the satellite states and when he decided to keep all the gains he had made in Poland. This attitude was reinforced over the next few years when he attempted to exert pressure on Greece to gain control of more territory (as we will see later).

… And They Were Capable of Trying

At the end of World War II in Europe, the Soviets had more soldiers deployed than anyone else (a ratio of about four to one) and more tanks than anyone else (a ratio of about two to one).

In fact, the perceived threat level was high enough that the British actually mapped out the opening stages of an outbreak of war with the Soviets, called Operation Unthinkable (kept secret until the late 1990s). There were also ongoing conversations between the British and the United States as to the unfolding of such a new war.

What's a Cold War?

Because these tensions never turned hot (a military term used for areas of open fighting), the phrase "cold war" was coined. The Cold War, which began as World War II ended, was a war of economic competition to see who could fund the largest buildup of military and development of the most advanced kinds of weapons. It was also a war of influence (sometimes referred to as a proxy war), in which the Western countries and the Soviet countries supported and encouraged differing factions within the militaries and the governments of various smaller players on the world stage. The Cold War's struggle of spy versus spy dwarfed all previous examples. The American Central Intelligence Agency (CIA) and the Soviet KGB, just two of the many agencies involved, employed enough people to populate a small country.

Now you might wonder how that was different than a lot of other periods of history, where different empires were competing for expansion through economic and political means and didn't break out into open warfare. The key difference here is twofold. For one thing, during no other

Official government policy in the USSR guaranteed that artwork promoted the strength and integrity of the workers.

period in history were the two nations at the center of the tension anywhere near the scale of the United States or the USSR with its satellites. The only other such cases that would even be comparable *did* break out into open warfare.

For another thing, the scale of destruction that was always kept near the boiling point was truly mind-boggling. The stockpiles of nuclear missiles made this standoff unlike anything that had come before—especially since developers, over the course of the Cold War, designed bombs that were almost a thousand times more powerful than the one dropped on Hiroshima.

"Cold War" is not very fitting when you think about it. It should have been called something like "the almost-boiling-over" war.

Setting the Stage and the Opening Scene

In this book, we will examine some of the first few scenes of the Cold War story. We will follow countries like the United States, Britain, and the USSR as the driving decision-makers in setting out the policies and decisions that would shape the Cold War. We will also follow certain key figures, like British Prime Minister Winston Churchill, Stalin, and US President Harry S. Truman.

As the story develops, we will see the important roles played by smaller nations, like France and Greece. The heads of state will be joined by some of the analysts and support staff, like secretaries of state and members of the **intelligence** community, who helped create the policies and perspectives that would drive their governments' approaches. Additionally, we will see how large organizations, like NATO and the Warsaw Pact, were creating one kind of tension and how that contrasted with the goals of the newly-formed United Nations.

Throughout this book, our central goal will be to understand how the war began and to do it in such a way that we can look back later and see how the shape and direction of the Cold War were already revealed in the first stage.

The Austro-Hungarian empire at its height as the major power in Europe

Crowds turn out to show their support for the Communist Revolution.

Communists saw the political and military actions by the West against the long backdrop of tension, conflict, and what felt like Western attempts to interfere and keep Russia down.

In the period between the big wars, a lot of the small nations passed through a rough period with the agricultural economies being taken over by dictators (like Poland) or by

In fact, it was partly to preserve its diplomatic position and partly to live up to its role as the protector to these smaller neighbors—the older brother, so to speak—that Russia was pulled into World War I. This move resulted in the Communist Revolution. Russia suffered incredible losses in the war, and it was no small insult that Russia was not included in the meeting at Versailles after the war to decide the fates of different countries. Some of the countries whose fates were being decided were the ones that had thrived under Russian influence and protection. (Serbia, for instance, had lost more than a tenth of its total population in the war.) France and Britain, the very nations that had made Russia let go of gains after its victory over the Ottomans, decided they should be the new owners of very valuable holdings such as Iraq and Syria.

Birth of the USSR

The Communist Revolution in Russia came hard on the heels of the massive loss of resources and lives during World War I. In the civil war that started in 1917 and would last until 1922, countries of the west backed the White Movement to try and resist the Communists. This force included troops from the United States, Britain (and its commonwealth nations), Japan, and France. Their efforts failed, and they had to withdraw by 1920.

However, Stalin would emerge in 1922 as the leader of the Soviet Union in the free-for-all fight that broke out after Vladimir Lenin, the previous leader, died. He and the

A second element that would come into play was similar strife with Japan over gains Russia made in China and over Japan's move to take control of Korea before Russia could. This would set the stage for additional conflict after World War II.

The third element that contributed to the war was the fact that part of Russia's efforts was not merely conquest. They also helped to establish the new nations of Serbia and Bulgaria, known as "little **Slavic** brothers" to Russia.

Russia and Britain gained additional influence in regions like Afghanistan and Persia, and they were in competition with each other in this race to empire. Their efforts managed to divide a lot of the geography between them. This competition was patched together in places with many uneasy truces.

With these problems, though, the many agriculture-based nations of Eastern Europe were industrializing much more slowly than Russia and the countries of Western Europe and across the Atlantic. The nineteenth century, though, was practically a **renaissance** of art and literature among these largely Slavic nations. Examples include the rise of the Russian novel under the skilled writing of figures like Leo Tolstoy. Following the lead of Russian composers like Pyotr Tchaikovsky, there were great composers like Frederic Chopin (Polish) and Franz Liszt (Hungarian). There were also great painters and, let us not forget, the Czech biologist, Gregor Mendel, who helped untangle the mystery of genetics or Ivan Pavlov, the Russian psychologist, who helped uncover the nature of trained behavior.

CHAPTER ONE

Mounting Tensions

It is fair to say that any major event or conflict is woven together from many threads reaching back into the past. In the case of understanding the Cold War world that would rise from the ashes of World War II, though, it is particularly helpful. By pointing out a few of the long-standing tensions between Russia and the West, it will help us to see the Russian perspective more fairly, instead of just the filtered version that evolved to validate the Western position during the Cold War. There are at least three elements that contributed to the tensions that eventually created enough impact for a war.

One element is the fact that Russia had managed to expand its territorial holdings during the nineteenth century. It did this to try to keep up with some of the other countries that pursued imperialist foreign policies—like Britain and France. In the 1870s, Russia gained a lot of territory after war with the Ottoman Empire, gains from which Britain and France pushed hard for them to withdraw.

monarchs seizing power (as in Yugoslavia). Social tension was increased by dropping prices in agriculture and a refusal to implement land reform. This pushed peasants to rebel, and the aristocrats ended up supporting the new authoritarian regime in order to maintain order. It is little surprise, then, that they had only faint resistance to offer when Germany rolled in and no viable structure of their own to restore once the Soviet Union pushed the Germans out.

Onward into World War II

After the policy of appeasing Adolf Hitler had failed, Britain and France switched gears and guaranteed the safety of Poland, which Germany was also threatening to invade. Stalin tried to join in forming a common response, but the other two seemed to distrust him almost as much as they distrusted the Nazis. This suspicion and their change in policy added fuel to Stalin's belief that they were trying to deflect Germany's military toward the Soviet Union. In the summer of 1939, then, this distrust led to a nonaggression treaty between Germany and the Soviet Union, even though their ideologies did not agree with each other. This was a big shock to the rest of the world. As you can imagine, it gave America and other countries a distrust of the Soviet Union to balance the Soviet Union's distrust of the West.

Not only did each agree not to attack the other, but they also agreed to stay neutral in case one of them attacked another country. The German-Soviet Commercial Agreement arranged for the Soviets to trade raw materials in exchange for German military equipment.

There was also a secret agreement (part of the Molotov-Ribbentrop Pact, named after the Soviet and German foreign ministers) that divided control of Eastern Europe between them. This allowed the Soviets to march into Finland and eastern Poland, reclaiming areas that had been taken from them in World War I. (As another result, Poland was split in two with part under German control and part under Soviet control.)

In September 1939, the USSR moved troops in to occupy Poland. Then, in June 1940, Latvia was occupied. Lithuania and Estonia were taken over on June 16, 1940. In late June 1940, northern Bukovina was occupied. Finally, in March 1940, eastern Finland was taken over.

The USSR relationship with the West went downhill a little more when the Soviets began building up their economic trade with Germany. Soviets sent things like oil, rubber, manganese (essential for iron and steel production) and so on to Germany. In exchange, the Germans sent them machines and technology for their factories to help with efforts at full-speed **industrialization**. They also sent weapons, which would come back to haunt them later. Naturally, this cozy arrangement came to a screeching halt when Germany short-circuited and decided to invade the Soviet Union. The invasion was called Operation Barbarossa.

In fact, the tension and mistrust were so high that even after Hitler broke the agreements and invaded the Soviet Union, he banked on the fact that the three major powers—Russia, Britain, and the United States—would not stay united in their war against him but would break up over disagreements. This mistrust was amplified because Stalin

Frank Wisner's Fight Against Communism

Mississippi lawyer Frank Wisner is known as the father of American covert operations. Bored with his new law career, he signed up for the military just months before Japan dragged the United States into World War II. He was assigned to the Office of Strategic Services (OSS), a **clandestine** intelligence agency. At the end of the war, he was hired on by a successful Wall Street law firm, and that might have been the end of his career as a spy.

However, because of his experience with OSS and his understanding of the Soviets, Wisner was recruited in 1947 into the State Department. He pushed hard for the creation of a new agency focused on **Communism**. Once it went through, it was up to Wisner to build the organization, focusing on areas like propaganda, sabotage, subversion, and the care and feeding of underground resistance groups.

One of his first big operations, Operation Mockingbird, was not focused on foreign countries but on influencing the American media. Another project was Operation Bloodstone, which recruited former German officers and **diplomats** that could help combat Communism. His other big success was to orchestrate the overthrow of the new president of Guatemala.

It was not all sunshine. He also had some failures, like his project to help achieve Hungarian independence. The plan imploded when the Soviets marched in, and the United States failed to give Wisner's plan military backup.

The CIA was a key organization behind the scenes of the Cold War from start to finish, and Frank Wisner was a key player within it.

felt like the others had waited and waited before opening the second front against Germany. He felt like they wanted Hitler to wear down the Soviet Union so they would be weaker, even after the war was over.

One of the overarching goals for the Soviets in World War II was to build up control over a bunch of smaller countries that would be a kind of buffer or shield against Germany. It was these efforts that caused tension in international relations, reflected in the conference at Yalta in February 1945. The same increasing strain was also felt at the Potsdam Conference a few months later in August.

Just as the Soviets were looking to restore their role and control over the other Slavic states, the countries of the West feared—as always—the power that would be wielded by a larger, stronger Soviet Union.

Yalta Conference

In 1945, the big three—President Franklin D. Roosevelt of the United States, Prime Minister Churchill of Britain, and Premier Stalin of the Soviet Union—met at Yalta, a small coastal town on the Crimean Peninsula in the southwest part of Russia, which sticks out into the Black Sea. This city on the coast had been occupied by a remarkable variety of empires down through history, including Romans, Greeks, Byzantines, and the Ottoman Empire. It had just been occupied by German forces for over three years until they were pushed out in 1944. It seemed to be a fitting location for the leaders of the most powerful allied nations to meet and discuss the fate of nations. Their goal was to discuss

Churchill, Roosevelt, and Stalin—the big three—at the conference in Yalta

Mounting Tensions

strategy and to coordinate plans for the final stage of war against Germany and for the occupation that would follow.

In a move that would have unfortunate repercussions over the next few decades, the president of France, Charles de Gaulle, was not invited to the Yalta Conference (nor the Potsdam Conference a few months later, in July 1945).

Churchill kept pushing to include France at these meetings, but Roosevelt pushed back (and slightly harder), arguing that giving France a seat at the table would just complicate things. However, in fairness, Roosevelt supported Churchill at the actual meeting in Yalta in fighting for France to be given control of one of the occupation zones in Germany. The US president also set France in position to have a permanent seat on the security council of the United Nations by having them be one of the countries that invited others to the conference.

Over the course of the conference, the leaders made key decisions about the dismantling of German military industries, formed a plan for trying key war criminals in a trial at Nuremburg, and discussed how war **reparations** would be handled. Perhaps the main focus of the meeting, though, was on how to restore the various defeated and conquered countries. In resolving this last element, the three reached agreements on setting up interim governments and arranging for democratic elections to take place as soon as they could. Any guesses on whether Stalin kept his part of the bargain and allowed democratic elections in places like Hungary and Bulgaria? Poland would prove to be the trickiest, with Stalin unwilling to cave in to demands on one

Cecilienhof Palace, site of the Potsdam Conference

side and the United States and Britain unwilling to cave on the other. Everyone was forced to settle for a compromise.

It should be mentioned that the group also cut a secret deal to give the Soviet Union some Japanese islands and some territory Russia lost in the 1904 war with Japan. In exchange, Stalin agreed to enter the war still being fought with Japan once the group finished crushing Hitler.

The Soviets and Art

The tension driving this suspicion and mistrust was not just a tension in politics but a tension in underlying visions as well. The Soviet view of human nature was a different one from that so prominent in the West. To the eyes of the Communist Party, Western art and culture reflected a kind of decadence and excess.

The Soviet government worked to promote a more focused social sphere of art and music oriented around something called socialist realism. This meant art should lift up workers, farmers, and soldiers—portray *those* as noble, heroic pursuits. Similarly, it should champion the ideas of equality, not entertain people with false stories of extravagant living.

When politician Andrey Zhdanov laid out the elements and goals of Soviet policy on art in 1934, one of the key ideas he included was that writers should see themselves as engineers of the soul. He felt they carried a heavy responsibility not to sink down into sappy romanticism or portray lives of lazy materialism as joyful.

The great writer Aleksandr Solzhenitsyn was kicked out of the Soviet Union because some of his writing cast the government in a negative light and challenged the values they were working to instill. Even Solzhenitsyn, however, was overwhelmed by the individualism and the materialism he saw when he came to the West.

Potsdam Conference

Potsdam is an old city just outside Berlin and the capital of the district of Brandenburg. In fact, the group met in the palace that had belonged to the family who ruled the German Empire up until the end of World War I, the Hohenzollerns. It was here that the big three Allies reconvened to push forward on ironing out the final borders of Poland, to set up the four occupied zones of Germany that they had planned at Yalta, and to divide Berlin, Vienna, and Austria into four zones in the same way. They also planned more of their strategy for the war with Japan.

This conference went from July 1945 into August 1945. President Roosevelt had died in April 1945, so the new president, Harry Truman, was now the third participant. Churchill lost the election during the conference, which meant the second part was attended by the new British Prime Minister, Clement Atlee. Stalin was still the premier of the Soviet Union this entire time.

In spite of the tensions and the fear of a kind of new Russian Empire, the Allies at the Potsdam Conference had to concede to letting the Soviets keep their gains in Finland, Germany, Poland, and Romania. They also kept power over the Balkans. The fear of giving the Soviets territories and resources like this caused some European countries such as Britain to sweat a little more because they thought the United States might go back to the policy of isolationism they had adopted before the war. Once United States troops withdrew from Europe, as Roosevelt stated at Yalta, it felt like the European states would be the ones who would have to face the greater pressure of expanded Soviet forces.

Looking Forward

During Yalta and Potsdam, we saw the collaboration among the three world powers that was to characterize the birth and role of the United Nations, but we should also notice the changing tensions between the Western Allies and the Soviet

This map shows the dividing line through Europe created by the Iron Curtain.

Union as each sought to exert control and gain as much power as they could. This growing distrust foreshadowed the dramatic **schism** that was soon to expand to full scale in the form of the Cold War.

Over the coming chapters, we look at how the conflict grew between nations holding loyalty to one of the two grand theories about human nature and about how to create the best society. In the USSR, there was an ongoing series of purges by the Communist Party to try and kill, or **exile** to cold Siberian prison camps, all those who disagreed with Stalin's narrow version of Communism. Stalin and others crushed countless lives under the wheels of this machine. In the United States, there was not as much killing and exile, but the rise of McCarthyism still had a devastating effect on lives and careers. The movement was named for Senator Joseph McCarthy, who was violently opposed to Communism and pursued it (with his supporters) with the passion and extremism of a witch hunt. This other dimension is the way in which the division and conflict between the two theories of human nature also happened inside the countries, not just between them.

In more populated areas, like Berlin, the Iron Curtain was stronger and manned by security personnel.

CHAPTER TWO

NATO, the Warsaw Pact, and the Iron Curtain

We have a solid overview now that we have walked through some of the tensions and trends leading up to the formal outbreak of what would come to be known as the Cold War. In this chapter, we will look at three of the key features of that outbreak: the Iron Curtain, NATO (North Atlantic Treaty Organization), and the Warsaw Pact. In particular, we will focus on the physical and **ideological** lines that were drawn and used to divide the world into two different camps. Through understanding the significance of these dividing lines, we will begin to see each nation's motivations and actions emerge, setting the stage for the Cold War.

Iron Curtain Speech

It is difficult to chop a river in pieces while it is moving, and history is like that in a lot of ways. However, if we want to understand the shape of the overall river, we can look at certain important points where it goes through rapids or makes a sharp turn. In understanding the series of developments that led to the creation of NATO and the Warsaw Pact, a famous speech by Winston Churchill is a good example of just such a bend in the river.

While it is often called the "Iron Curtain Speech," its formal title was "Sinews of Peace."

Churchill had been invited over by President Truman in order to make a public statement to the world, so to speak, of the view shared by Britain, the United States, and the other Western nations. The pretext for the visit and the speech was that Churchill was being awarded an honorary doctorate from a little college in Missouri called Westminster College. On March 5, 1946, Churchill introduced us to the term

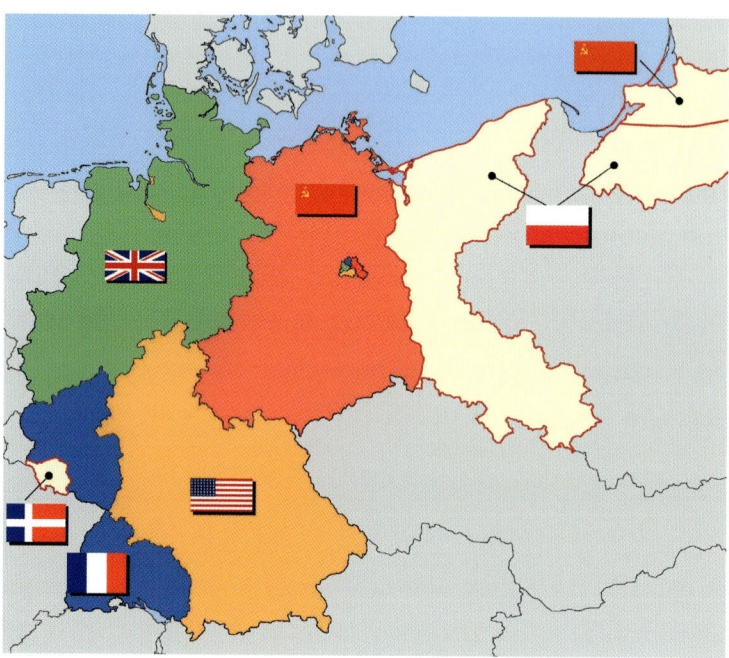

This map shows Germany's division into separate occupied zones.

28   NATO, the Warsaw Pact, and the Iron Curtain

"Iron Curtain," which would become a buzzword during the Cold War. He also famously pinned down what people thought of as the geographic endpoints with this famous quote: "From Stettin in the Baltic to Trieste in the Adriatic, an 'iron curtain' has descended across the continent." (Stettin is a city in Poland, and Trieste is a city in the northeast part of Italy.)

The speech was not very popular after it was given, and a lot of people in the Western countries were still used to thinking of the Soviets as being on their team in World War II, working against the Nazis and the Japanese. Over time, though, as more people came to understand the new realities and tensions taking shape in Europe, the speech and the metaphor itself became more accepted.

Churchill had been defeated in the recent British elections, so he was speaking as a private citizen, not as an official head of state. Because of that, though, he could speak more openly about the growing distrust of the collection of Communist countries.

In his speech, Churchill offered some strong criticism of Soviet actions and policies in taking control of various smaller countries in Eastern Europe. He also gave some **foreshadowing** of developments that would be coming in the near future. This included the need for stronger cooperation between the United States and Britain, along with others, to keep the Soviet Union from continuing to expand their power.

Not long after this, Stalin gave his response in *Pravda*, the Soviet newspaper. In his view, Churchill and the West were trying to mischaracterize the friendship among the

Stalin Responds to Churchill

It is important to understand how others imagine themselves to be the good side. Here we see a part of Stalin's response to Churchill in an interview with *Pravda* (the Soviet newspaper) published in the *New York Times:*

> As a result of the German invasion, the Soviet Union has irrevocably lost in battles with the Germans, and also during the German occupation and through the expulsion of Soviet citizens to German slave labor camps, about 7,000,000 people. In other words, the Soviet Union has lost in men several times more than Britain and the United States together.
>
> It may be that some quarters are trying to push into oblivion these sacrifices of the Soviet people which insured the liberation of Europe from the Hitlerite yoke.
>
> But the Soviet Union cannot forget them. One can ask therefore, what can be surprising in the fact that the Soviet Union, in a desire to ensure its security for the future, tries to achieve that these countries should have governments whose relations to the Soviet Union are loyal? How can one, without having lost one's reason, qualify these peaceful aspirations of the Soviet Union as "expansionist tendencies" of our Government? ...
>
> Mr. Churchill wanders around the truth when he speaks of the growth of the influence of the Communist parties in Eastern Europe. ... The influence of the Communists grew because during the hard years of the mastery of fascism in Europe, Communists slowed themselves to be reliable, daring and self-sacrificing fighters against fascist regimes for the liberty of peoples.

Soviet Union and the smaller states around it. He claimed that the accusations of seeking to expand their power were just warmongering—an excuse to continue working to keep Russia and the others down and to use force to spread their own power and agenda. In an interesting (and clever) twist, Stalin claimed it was the West that was imposing an "iron curtain" to keep information about the success of Communism and Soviet cultural achievements from leaking out.

The Geography

With the Iron Curtain in place, preventing information or people from passing through, the Soviet Union had control over:

- Estonia
- Latvia
- Lithuania
- Part of eastern Poland
- Part of eastern Finland
- Northern section of Romania
- Northern half of what used to be east Prussia

These areas were merged into Soviet districts, so they would no longer show up on a map of Europe at that time. In addition to the ones it more or less swallowed, the Soviet Union also gained control of the following states (called satellite states afterward) in the years 1945–1949:

- The German Democratic Republic
- The People's Republic of Bulgaria

- The People's Republic of Poland
- The People's Republic of Hungary
- The Czechoslovak Socialist Republic
- The People's Republic of Romania
- The People's Republic of Albania (which re-aligned itself in the 1960s away from the Soviet Union and toward the People's Republic of China)

The Soviet Union installed people in government that would be loyal to them and follow directions. The only exception was Yugoslavia, which managed to retain its own independence.

Outside the Iron Curtain, then, were the Western nations. Many of these states began working to form alliances with the United States. This included most of Europe with a few exceptions (Austria, Finland, Liechtenstein, Malta, Republic of Ireland, Sweden, and Switzerland). NATO would eventually rise from this. While there seem to be a lot of states that did not rush to form alliances with the United States, they were still market economies and governments guided by democratic practices and principles. Therefore, they were more closely connected with the United States than with the countries to the east.

The Berlin Airlift

Truman sent administration officials in 1947 to work with the Soviet foreign minister Vyacheslav Molotov about how they could reach an agreement on working to create a stable

German economy and help it stand on its own two feet. He thought that without a strong, stable Germany, there would not be a strong, stable Europe.

Sadly, the skepticism of Molotov was justified—Russia has been invaded three different times in the last 150 years and lost millions and millions of people, either through poor conditions or through these wars of invaders. As a result, in high Soviet fashion (during the sixth week of negotiations), Molotov refused the plan and adjourned the talks.

With the dreams of the Third Reich in rubble-strewn ruins, the major powers had each established control of different sections of Berlin. In many ways, this would serve as a microcosm of the larger conflicts about to reshape the world stage. In 1948, the United States, Britain, and France decided to merge their zones into one economic group with the German mark as their shared currency. Of course, this led to much contention with the Soviet Union. The Soviet Union argued that this economic action by the Western powers broke the rules of some of the earlier agreements and declared that therefore the Allies no longer had any jurisdiction in Berlin.

To retaliate against the West, the Soviet Union closed off all the roads, railways, and waterways that would allow goods and raw materials to pass in and out of west Berlin. Within two days, the United States and Britain began bringing goods in by air. In an impressive display of power, they kept this up for eleven months around the clock. Not only was the Soviet Union not able to stop them from flying in the goods, but the Western countries also put an **embargo** on all strategic goods coming out of the countries behind the curtain. The

The Berlin Airlift was one of the largest and longest relief efforts ever accomplished.

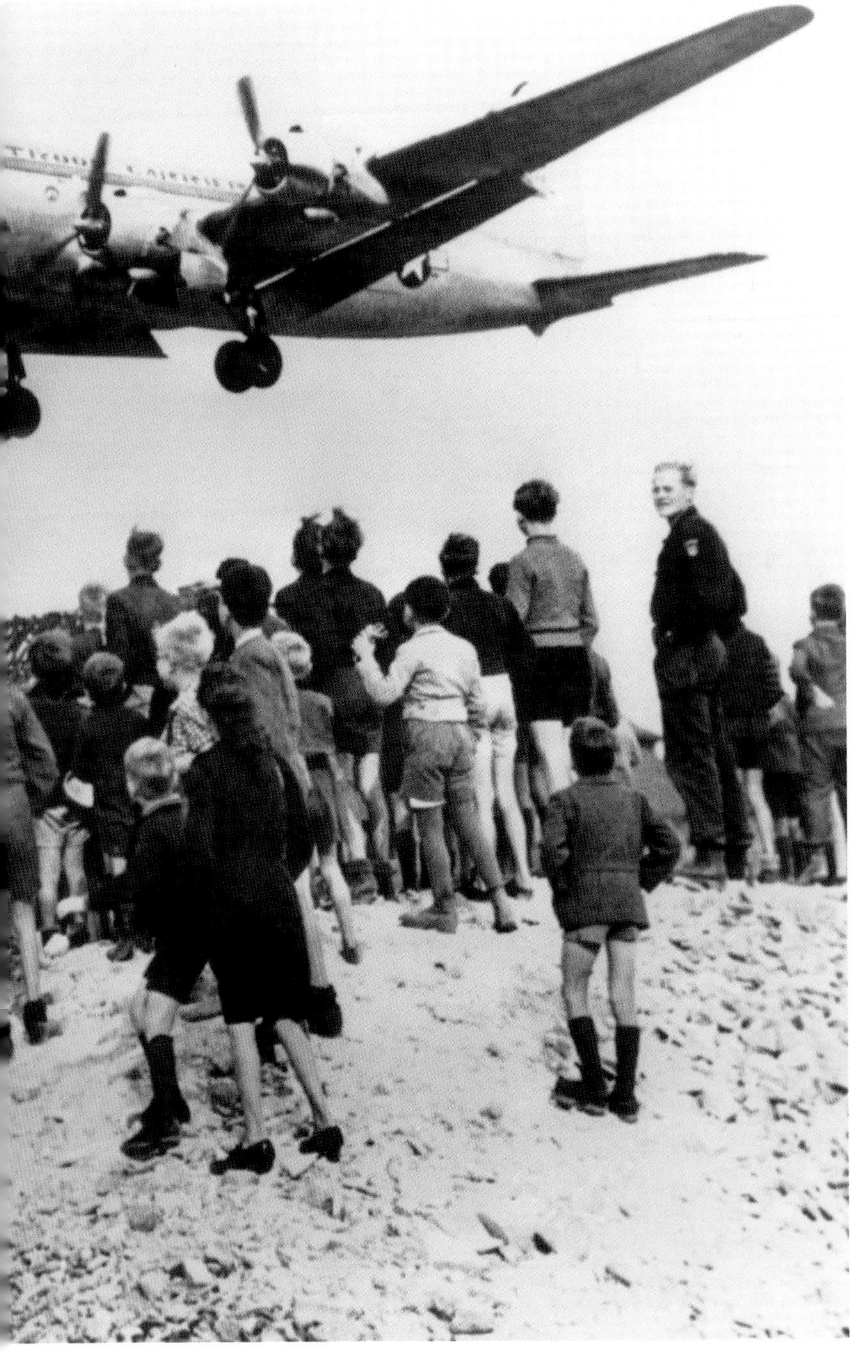

Soviet Union finally called off the blockade in 1949. By the end, the airlift had taken fifteen months. There were slightly more than 277,000 flights into Berlin, and the airlift had delivered over two million tons (1.8 million metric tons) of machines, food, and so on. They even had to fly in the coal that the Berlin residents needed to heat their homes. If this had not worked, there was a backup plan in place to take an armed convoy through the section of Germany that the Soviet controlled. It might well have been World War III just waiting to happen.

The merged sector formed by the Western countries was formally organized to become the Federal Republic of Germany (FRG), known as West Germany. Not to be outdone, the Soviet Union turned their occupied zone into the German Democratic Republic (GDR) in October of that same year.

Restricted Emigration

Having learned important lessons at the end of World War I, the Soviet Union ended World War II with forces occupying a lot of the neighboring Slavic states. Bargaining from a position of strength, they were able to negotiate and regain of much what was stripped away from them in 1918.

As discussed earlier, Stalin wanted to make sure all of Germany's military teeth had been pulled out by completely dismantling their industrial production capacity. Britain, however, saw the potential value of a restored Germany as an ally against Soviet expansion—a point of view that did not go unnoticed.

By the next year, 1950, migration from inside the Iron Curtain was more or less closed down, though they did still allow it in certain operations. This was a sharp contrast from the more than 15 million people who left Soviet-occupied countries to flee to the west from 1945–1950. Once the clampdown happened, there were only about thirteen million for the next forty years!

The Actual Wall Itself

Now let's look at the events that helped create a physical version of the ideological divide between the different

In less populated areas, like this rural scene, the Iron Curtain was often less fortified.

Escaping the Wall

The goal of the Berlin Wall was to prevent people from escaping because the loss of too many educated people and skilled workers would leave holes in the East German economy. (How do you run a city if all the doctors and auto mechanics have left?) No system is perfect, of course, and the system in place to try and stop people was no exception.

How many people escaped? Estimates say that around five thousand breakouts happened over the years. These escapes showed different levels of difficulty and different levels of ingenuity.

In the early days, officials had not yet sealed up the windows of all the buildings next to the wall, so some people got away by jumping out the window on a high-enough floor of their building. Other straightforward strategies sometimes worked as well, like smashing against the wall with a big truck and then racing for freedom.

Next up on the scale are things that might be easy to think of, but hard to pull off, like digging a tunnel from a basement far enough to go under the wall. There is even video footage of one group who made it through the wire fence and guards on the eastern side of the wall and still had to swim across the river faster than the patrol boat could respond to their location. Maybe the most intriguing, though, was the group who sewed together saved pieces of cloth to make a hot air balloon.

groups of countries. Naturally enough, the border between West Germany (Federal Republic of Germany) and East Germany (German Democratic Republic) was one of the places where the Iron Curtain took shape in the most obvious and oppressive ways. It could be seen cutting through the rural parts of the country in the form of sharp, steel mesh fences. In the urban parts of the country, it took a more formidable shape—often as a tall concrete wall.

A great example of this, and surely the most famous part of the wall, was the Berlin Wall itself, which was installed in 1961. At its greatest length in the 1980s, it stretched about 28 miles (45 kilometers) through Berlin and another 75 miles (120.7 kilometers) around the edge of the city. This was a concrete wall about 15 feet (4.6 meters) high, with razor-sharp barbed wire on the top. There were also signs all along the wall, and there were guards posted at areas with gates. Since it was located somewhat inside the actual border, the East Germans were able to have patrols guarding the strip on the West German side of the wall. Sensibly enough, there were also West German guards patrolling along the border area at the edge of the strip to help safeguard their own citizens. Over the years, additional features were added along the East German side, including anti-vehicle ditches (so cars and trucks would get stuck), electric fencing, and a minefield.

It was probably the watch towers looming over the landscape at regular intervals that most helped change the feeling of the Berlin Wall from a mere barrier to something more like a prison wall. In a way, it *was* a prison wall. A key motive for building it was that there were so many educated professionals and skilled workers leaving East Germany that it was causing harm to their economy. Over the years, about five thousand people escaped over the wall, but about the same number were captured while trying, and close to two hundred people were actually killed.

In other places, the Iron Curtain took various forms. In Hungary, as an example, the border zone started 9.3 miles (15 kilometers) away from the edge. There was no giant concrete wall, but it had an area just over 50 yards (45.7 meters) wide that was full of landmines, and there were high fences on each side of it. (Later, they took out the landmines, switched to electrical fences, and added guard towers.) People would need special permissions to even enter the area—like if they lived or worked closer to the border than the 9.3-mile (15 km) limit. On top of all that, there were guards patrolling around the clock, and they were allowed to shoot people trying to escape.

The physical dividing line created by the Iron Curtain was one thing, but even more important was that it also represented the ideological divide with an array of countries lined up on each side. Their struggles would cause the dividing line to extend much further, reaching into the airwaves, for instance, as Radio Free Europe—a radio broadcasting organization created by the US government in 1950—worked to broadcast news and information into the countries locked behind the wall. They also extended the curtain into a race to develop the most powerful nuclear weapons and the fastest computers. It eventually became a competition to explore space—a race that continues even to this day.

NATO

The European countries, as already reflected in some of the discussions at Yalta and Potsdam, knew that there would need to be a collaborative organization to put military force behind their ideas and their intention to prevent the spread of Communism and of the Soviet Union. Countries working to rebuild experienced a tangible fear that conflict with the

Representatives at the first NATO conference

Soviet Union and its agenda could easily drag the world right into World War III with hardly enough time to even gas up the planes and tighten up the tank treads (maybe a tiny bit of exaggeration but not by much). Several regional treaties were organized, like the Rio Treaty (1947), for countries of the Western Hemisphere, and the Brussels Treaty (1948), signed between France, Britain, and the Benelux countries (Belgium, Netherlands, and Luxembourg).

However, since the Soviet hammer would fall on Europe first, leaders in both Europe and the United States realized that an organization that added the United States to the European alliance agreements would be necessary to deter Soviet aggression.

The United States, Britain, and Canada discussed moving toward this end for a good while. In fact, the signing of the Brussels Treaty was partially a result of America telling the European countries that they needed to show commitment to the idea of mutual alliances before it would consider being tied to their defense.

After back-and-forth rounds of negotiation that seemed unending at times, the North Atlantic Treaty was signed April 4, 1949. The original signers of the treaty represented:

- Belgium
- Canada
- Denmark
- France
- Iceland
- Italy

- Luxembourg
- the Netherlands
- Norway
- Portugal
- the United Kingdom (Britain)
- the United States

Other countries were admitted early on, such as Greece and Turkey in 1952 and West Germany in 1955. Spain joined in 1982. Many countries were admitted in the 1990s and early 2000s, as the Cold War came to an end—mostly.

The treaty itself was intentionally short and written in language straightforward enough that everyday people would be able to read and understand what it was saying. There are fourteen different articles (parts of the treaty), with each setting out some specific commitment. For instance, Article 3 covers the commitment for each country to maintain its ability to resist attack. In other words, it had to keep up a military of its own. In Article 4, member countries agree to consult with the others when one of them is threatened. Article 8 is an interesting case: the countries agree that they are not tied up in any other agreements or alliances that would conflict with this one, and that they won't agree to one later.

Of all the negotiating that went into the document, the hardest piece to get right was Article 5, which commits the members to act if one of the other members is attacked. Perhaps because the members had calculated so well in designing the document and the alliances, the first time

The attack on the World Trade Center in 2001 prompted a rare response from NATO.

NATO, the Warsaw Pact, and the Iron Curtain

they needed to invoke Article 5 was in 2001 when terrorists attacked the World Trade Center and the Pentagon in the United States.

Warsaw Pact

One of the keys to rebuilding Europe was rebuilding Germany and putting it on a stable footing. Of course, the main way to accomplish that—rendering it stable—was to figure out a way to provide for German security without creating fears of armed aggression. One solution attempted was to organize a collective force, called the European Defense Community (EDC). This would have been a separate army, not wholly under the command of one country or the others. After some back and forth attempts to negotiate, the French voted down the creation of the EDC.

In response, an alternative solution was brought into being. That solution was to admit West Germany to the NATO alliance. Rather than give Germany control over such a large military, the German forces were put under the command of the Allies.

This, as you might guess, went against the Soviet desire to keep Germany a weakened state. Seeing, instead, a new and strong Germany and a boost to the military strength exercised by NATO, the Soviet Union formed a counterbalancing organization known as the Warsaw Pact. Established in July 1955, the official name of the document (translated to English) was the Warsaw Treaty of Friendship, Cooperation, and Mutual Assistance. In addition to creating a counterbalance to these particular moves by the West, it also allowed the Soviet Union to exert more military control over the small satellite nations.

The nations signed into the Soviet agreement were:

- Albania
- Bulgaria
- Czech
- Slovakia
- German (East)
- Hungarian
- Poland
- Romania
- Russia

Given the nature of the centralized control Russia used in decision-making, the Warsaw Pact obviously did not have the kind of agreement or consent for mutual aid that NATO had. Its secondary purpose was, in fact, to be used as a tool to sweep in and obliterate individuals or groups who are going against core standards of socialist doctrine or pushing back on various party directives.

Of course, this description is a view of the document and the organization from a perspective of someone from the West or even, as much as we can, looking at it from the viewpoint of an outside investigator. If you think about the things discussed in the first chapter, and after you read some of the discussion and the sidebars (here and in future chapters) it should be easy enough for you to imagine how differently the Soviets themselves would probably characterize the agreement.

Joseph Stalin was in many ways both the hero and the villain of the Soviet rise to power.

We must be careful, however, in the process of engaging in such an attempt to be fair by looking at the situation and events through the eyes of the different major figures involved. In particular, we must be careful not to fall into thinking that the view from those different perspectives is equally correct. Consider for instance that there are hard, cold facts about the case that will pop the bubble being created sometimes in the midst of passionate defense. Here we might well ask if it is coherent or strained for the Soviets to claim they are just working to protect their people from the contamination by Western ideas when they killed hundreds of thousands of them to smooth things out. How many people can you send off to die in frozen, Siberian work camps and still claim you are being a "big brother?"

In fact, historians often point out that Stalin might actually be worse than Hitler with respect to the scale of intentional murders he ordered or caused. One of the largest acts of destruction was something called the Great Purge, also known as the Great Terror. It is estimated that this series of actions from 1936–1938 was responsible for killing as many as a million people directly and sending nearly ten million to the gulags, the horrible prison camps in the cold northern part of Russia. For countless Soviets, the camps were a slow death sentence. Many were members of the Communist Party or the secret police that Stalin considered to be dangerous threats to his tight control.

At the same time, though, there was also widespread starvation and death under his iron-fisted policies of land reform (mentioned earlier in chapter one). Stalin had around thirty thousand landowners killed and marched another two million of them off to the work camps in order to proceed with his plans. This then set off a grain shortage and famine, which killed millions of peasants. Because there was such limited record keeping, it is hard to say exactly how many died.

The ideas discussed in this chapter can be kept in mind when learning about some of the figures and political machinery that set these actions into motion, in the next chapter.

President Franklin D. Roosevelt helped set the stage for the Cold War, but he would die before it was officially under way.

CHAPTER THREE

The Faces of the Cold War

With a broad picture in mind of the stage setting and the major themes that seem to be at work in the unfolding story of the Cold War, here is a short study of some of the key characters. They should provide a feel not only for important turning points, but also for the different types of people involved. On the one hand, there are the center-stage world leaders, like presidents and prime ministers. On the other hand, there are also the behind-the-scenes figures— like political advisors and policy analysts in the intelligence community—who sometimes do the larger portion of the work.

Joseph Stalin

As much as we can say any one figure stands at the center of a complex series of historical events, then Stalin is such a man. It is his policies, his personality, and his perspective against which the other characters stood and the numerous decisions were made.

Stalin was born with the name Ioseb Dzhugashvili (pronounced "Joo' gosh yee' lee"), in Gori, Georgia (Russia), on December 6, 1879. He died March 5, 1953, in Moscow. His highest position was premier of the Soviet Union (1941–1953). He had also been secretary general for the Communist Party from 1922 until 1953.

During the course of his career, his accomplishments included a full-speed industrialization of agriculture, expansion of the territorial holdings of the Soviet Union, and cooperation in defeating Germany in World War II.

Stalin's father was a cobbler in their small rural town, and his mother was a maid. His father was also an alcoholic who used to beat him. He attended a religious school and went on to study at Tiflis Seminary. He read the work of German philosopher Karl Marx there, which was not permitted, and he ended up being expelled in 1899 for stirring up ideas about revolution.

In 1900, he was active in politics, specializing in things like agitating to create labor demonstrations and fueling confrontations with the police. It is little surprise that he went on, in 1903, to join the most militant of Communism's two main factions, the **Bolsheviks**. He studied under Lenin, the leader of the group. In the space of about ten years, he managed to land in jail seven times. As a revolutionary, that was a sign of his passion and active engagement, which helps us to appreciate how he proceeded to move up the party hierarchy again and again.

month of being in there, Churchill escaped and headed back to England. Returning as a larger-than-life hero, he won his next election, and his long, impactful career in politics was soon under way.

Over the course of the next sixty-five years, from his first election in 1900, Churchill would be in and out of office, moving ever upward. His experiences with soldiering in World War I also provided him with valuable experience and insights regarding war with Germany that would serve him well the next time around.

There is a central strength in his way of seeing that stands out and shows the perspective he was able to bring to the table. That strength was his ability to stay focused on the long game. For instance, he was able to put aside his opposition to Communism. It would have gotten in the way of the alliance needed to beat Hitler. He brought his passionate stance against Communism back into his decision processes once the job was done. A second great example of this was his dogged persistence. He lost important battles in World War I, he lost elections, and he lost popular support at times. He kept at each project, though, and that often led to outstanding results.

Charles de Gaulle

De Gaulle was the president of France, a decorated soldier (World War I) and military general, and the leader of the French Resistance during World War II.

Charles de Gaulle was born in Lille, France, on November 22, 1890, and died in November, 1970, at his home in Colombey,

Charles de Gaulle was an iconic member of the French Resistance during the war and a strong political leader afterward.

France. His family was devoutly Catholic, and his father was a history professor at a nearby Jesuit university. There are a couple of interesting things about the family that we can see influencing de Gaulle later on. The first is that his father liked to encourage good intellectual debate at the dinner table, and the second is that the family had lost a lot of their land in the French Revolution. These helped to shape both his ideological independence and, it is fair to say, his vision of France as standing large on the world stage, having reclaimed a status so well-deserved and so unfairly taken by German aggression.

After graduating from France's top military academy, he became a second lieutenant and was decorated for his efforts in the Battle of Verdun. He taught for a year back at Saint-Cyr, the military academy, and then after a two-year special training program, he was promoted to the Supreme War Council and later to National Defense Council. It was in World War II, as the leader of the Free French—which he organized—that his confident independence started to shine. He eventually managed to unify most of the French resistance groups under his leadership. One interesting fact: the commander he served under for Verdun was the man who was put in charge of the government for a time, so de Gaulle had to leave France and conduct resistance fighting from Algiers!

A key part of his positions and perspectives in the Cold War was his **advocacy** for France to adopt a "politics of **grandeur**." According to this approach, France should not be dependent on anyone else. In fact, this is central to his decision to withdraw from NATO. It also helps explain why he worked so hard to help France develop a nuclear research program, with the result that they became the fourth country to have nuclear weapons.

It is also well-known that he carried a grudge for the rest of his life over the fact that he was not invited to the conferences at Potsdam or Yalta. The man was persistent, so it's understandable. After all, we're talking about the same man who tried *five* times to escape when he was being held as a prisoner of war.

Harry S. Truman

It's hard not to like President Harry Truman. He served two terms between 1945 and 1953 and stands as a great example of how an average person can step up and do amazing things that change the direction of the world. Truman had tried farming and more or less failed at it. He had tried running a business and more or less failed at it. But once he was sitting in the chair behind that presidential desk …

Truman, the son of a farmer and mule trader, was born May 8, 1884, in Lamar, Missouri, a little farm town. Before he died in Independence, Missouri on December 26, 1972, he would manage to put the finishing touches on saving Europe from being overrun by the Germans. He would save the United States from being overrun by the Japanese and Germany from being swallowed up by the Soviets as well as help create the United Nations and navigate the new landscape of the Cold War's polarized world.

Unlike some of the other figures profiled here, Truman's work career was really all over the map: he was at different

President Harry S. Truman oversaw the close of World War II and the beginning stages of the Cold War.

The Faces of the Cold War

times a farmer, a bank clerk, a timekeeper on the railroads, and a soldier.

He had done two tours in the National Guard, so when World War I came around, he joined up. As a field artillery captain, he saw action in France. Stories indicate that he had a real talent for introducing order and discipline. His men appreciated his fairness and looked up to him for his bravery.

After returning from war and losing his shirt running a hat store, he made connections with the people who ran politics in Kansas City, and in 1922, he launched his political career: first local judge, then on to the Senate. He then won a second term, all along acquiring a reputation for his integrity and his flair for financial auditing.

In 1944, he ran as the vice president under Roosevelt, and they won. Strangely enough, he was only vice president for about eighty days before Roosevelt passed away unexpectedly. He was not well prepared for the larger role. In fact, he had only even met with Roosevelt a couple times. No one, including Roosevelt, knew there was a need to hurry. Truman stepped up to the challenge.

He was sworn in on April 12, 1945, and within months, he was sitting down with Churchill and Stalin at Potsdam. From that same conference, he would send Japan an ultimatum. In the early part of 1946, he had brought Churchill to Fulton to send out a warning cry about the dangers of additional expansion by the Soviets.

George C. Marshall

It says a lot about a person's ability when his former boss asks him to come back out of retirement to manage an important

General George C. Marshall helped author the plan that would guide United States policy in the Cold War.

project. In this case the boss was President Truman, and the project was helping prepare the United States military to land on Korea and push the Communists back. It says a lot about a person's character when he answers such a call. This kind of invitation actually begins the story of Marshall's career at the end.

George C. Marshall was born in Uniontown, Pennsylvania on December 31, 1880, and he died on October 16, 1959, in Washington. His mother and father were from families that had lived in Virginia since before the American Revolution. His father was a successful coal merchant.

He attended Virginia Military Institute, and when he graduated, he was commissioned a second lieutenant. His first posting was to the Philippines, and from then on his storied career reads like a fantasy novel of the ideal military man. With a great sense of responsibility and focus, it seemed natural that his path led to positions like general of the army, chief of staff for the army, secretary of state, and secretary of defense.

Marshall was sworn in as chief of staff of the army on the same day that Germany rolled its tanks into Poland. Starting with United States armed forces totaling 200,000, he had managed its growth in four years to a size of 8.3 million. (That's forty times larger for those of you keeping score.)

Churchill himself nicknamed Marshall "organizer of victory" for his leadership and ability in World War II. His contributions went well beyond the war, though. Marshall was also responsible for, of course, the Marshall Plan. This plan was central to European recovery after the end of World War II, and he drove discussions and was the key decision-maker for the treaty and formation of NATO.

In 1947, Truman called Marshall "the greatest living American" and appointed him as secretary of state. Moments like that are frosting on the cupcake for a remarkable career. The Marshall Plan was such a success that he was awarded the Nobel Peace Prize for it in 1953.

George F. Kennan

George Kennan was born in Milwaukee, Wisconsin, on February 16, 1904. His father was a tax lawyer, and his mother died a few months after he was born due to a ruptured appendix. He lived with his grandmother in Germany starting when he was eight, and he became fluent in what would be the first of many additional languages for him. He died on March 17, 2005, at his home in Princeton, New Jersey.

The rise of the CIA as a central player in the Cold War was largely due to the insights of former diplomat George Kennan.

The Georgetown Set

Both George Kennan and Frank Wisner were members of a social group called the Georgetown Set. It had originally started with a handful of former OSS members but grew to include a mixture of diplomats, journalists, politicians, and senior people from the CIA. They shared certain views, central among which was a passionate anti-Communist position.

Georgetown's cobbled streets lined with red brick houses were home to an elite social group called the Georgetown Set.

Wisner recruited some of the other members of the group into the CIA. When J. Edgar Hoover became jealous over the CIA's growing power, he started digging into the past and found that some of them had been active in left wing politics in the 1930s, so he passed this information to Joseph McCarthy. When McCarthy went after one of them, he got in over his head. The group turned the media machine they had created on McCarthy. His reputation was crushed beneath its wheels, and he never fully recovered.

This small circle of friends and associates, meeting over dinner and cocktails among the red brick houses and cobbled streets of one of Washington, DC's oldest neighborhoods, would shape the strategy and tactics for dealing with the Soviets. This same group would go on to advise presidents all the way up to Kennedy and Johnson.

Being a part of this group was a function of merit. Although many of the members of this circle had known some of the others at school or through political activities years earlier, these were some of the greatest experts in their fields.

Kennan attended Saint John's Military Academy in Delafield (near Milwaukee) and went on to study history at Princeton University. In 1926, he was hired on by the Foreign Service. His first post was in Geneva, Switzerland, but early on, he started receiving post assignments that would help build his knowledge of the Soviets. Kennan also attended an advanced study program on Russian culture, history, and language. Then, when Roosevelt opened diplomatic relations with them in 1933, Kennan was one of the diplomats assigned to Moscow.

It was after the war, though, when he returned to the United States, that he really had the opportunity to pull together the things he had learned. Marshall appointed him as the director of the policy planning group with the state department in 1947. Here, he was able to develop his ideas on the policy of containment. The key features of this policy had been spelled out in a 1946 communication that was known as the "long telegram."

This "long telegram" was actually a series of five telegraphs that walked through the elements of a policy of containment—by economic and technical resources when possible and by force when not. It also offered some of the theoretical observations about the Soviet mindset. This was an important piece of work and helped shape the United States and, through their influence, NATO's approach to the Soviet Union as the Cold War unfolded.

Interestingly, Kennan was also a strong supporter of the Central Intelligence Agency (CIA), which had been created in 1947, and he was the one who drafted the directive creating the Office of Special Projects in 1948. It was renamed as the Office of Policy Coordination and became the branch of the CIA concerned with **espionage**.

Kennan went on to serve as **ambassador** to the Soviet Union under President Truman and then as ambassador to Yugoslavia under President Kennedy. He finished out his career as an academic and a foreign policy critic at the Institute for Advanced Study in Princeton. As part of his work there, he authored a large pile of articles and a small pile of books—some of these won Pulitzer Prizes, National Book Awards, and other accolades.

Vyacheslav Mikhaylovich Molotov

Molotov, the son of a shop clerk, from a little village that was miles from anywhere, would rise to become the foreign minister for the Soviet Union and the advisor and right-hand man for Stalin for many years.

Vyacheslav Mikhaylovich Skyrabin was born on February 25, 1890, in Kukarka (now Sovietsk), a village about four hundred miles northeast of Moscow. He actually gave himself the name "Molotov" (from the Russian word for "hammer") after he became actively interested in politics.

Having been arrested several times for revolutionary activities, he was living in exile when he met Lenin. That led him to St. Petersburg to help with the revolutionaries' newspaper. From there, he went to Moscow. He was arrested again, and this time he was exiled to Siberia in 1915, but he escaped two years later and went back to St. Petersburg (which had been renamed Petrograd) with the revolution.

Vyacheslav Molotov was a key figure in the Communist Party, and his support was key in Stalin's rise to power.

With a talent for organizing and supporting revolutionary activities and incidents, Molotov kept working his way up in the party. Then, in 1926, when Lenin died, Molotov threw his support behind Stalin, and their collaboration from there would carry him to the top of the communist food chain.

Aside from his particular positions in the party—Prime Minister, Commissar of Foreign Affairs, and so on—he played an important role in several key events. The first

Berlin was in physical and economic ruin at the close of World War II.

of these is to note that Molotov was at the side of Stalin, listening and advising, at Yalta and Potsdam. The second is to consider the things he helped create. He was (obviously) the Molotov whose name was on the Molotov-Ribbentrop Pact with Germany. He also worked on alliances that were a precursor to the Warsaw Pact, and he had a hand in the formation and shaping of the United Nations. Churchill, having met Molotov many times, characterized him as a man of outstanding ability but also of cold-blooded ruthlessness.

Molotov died in Moscow on November 8, 1986. That means Molotov had worked with the men who stood at the sunrise of the dream of the Soviet Union and, with Mikhail Gorbachev (who launched his *perestroika* reforms that year), had watched the man who would oversee its sunset.

The United Nations Secretariat Building in New York was finished in 1952 as a beacon of hope and international cooperation.

CHAPTER FOUR

Negotiation and Innovation

Europe was pretty much a mess at the close of the war. During those first years, there was certainly promising economic growth. (America supplied much of the funding and materials). At the same time, though, there was a cloud of fear hanging over the region as everyone scrambled to outguess the Soviet Union and **hedge** their bets accordingly. There was also sadness because of the many lives and landscapes that were damaged or lost altogether.

Strangely enough, though, the level of damage turned out to be helpful in some ways. Since factories needed to be rebuilt, the manufacturing sector benefited. The plans and policies discussed in this section show how they were able to beat their old production levels and how, by 1955, economies began to grow.

Recovering in the Aftermath of War

The countries behind the curtain were not faring as well. They were also growing but not as quickly as the countries on the west side of the wall. On top of that, oftentimes, when one of its satellite countries had good earnings, the Soviet Union would siphon away a lot of that profit. Those funds then could not be poured back into the farm or business (or community).

Even at the same time they were working to entrench their hold on the competing spheres of the new polarized world, the superpowers came together and agreed on the creation of the United Nations, an organization aimed at maintaining world peace. The commitment to create such an organization was sparked and kindled among the allies from World War II. In 1944, most of the elements were put together by the United States, the Soviet Union, China, and Great Britain. Then in 1945, at a conference in San Francisco, the rest of the details were filled in by a larger group of representatives from fifty different countries.

The fact that both the United States and the Soviet Union were permanent members of the security council—each with veto power over proposals—meant that the United Nations would still always operate in the larger shadow of the Cold War. Nonetheless, with its intended focus on preserving the peace and on promoting friendly cooperation among the different nations, the United Nations was an important counterbalance to the divisive pressures of the Cold War.

Turkey and Greece

In 1946, Turkey became a testing ground when the Soviets moved into place to take control of the straits between the Mediterranean Sea and the Black Sea. Gaining better naval outlets had long been a dream of the Soviets and of Russia before that. Since the end of the war, the Soviet Union had been leaning on Turkey to let their ships pass through the straits between the Black Sea and the Mediterranean. Turkey was not favorably disposed to the plan, so the Soviets moved naval ships into the area to create additional pressure. Seeing the need to support Turkey, US President Truman sent an aircraft carrier, the USS *Franklin D. Roosevelt*. As hoped, this tilted the balance of power in the region, and the USSR backed down rather than risk going up against the greater firepower of the US Navy.

Struggle for control of the Dardanelles, in the northwest region of Turkey near Greece, almost sparked a new war.

Negotiation and Innovation 75

Then in 1947, the British said they could no longer carry on the effort of keeping the Greek **monarchy** in power against Communist forces. The United States felt that, given the tensions between Greece and Turkey in the region, neither could be left unsupported. If the Soviets got Turkey to change their minds and gained commercial and military access to the Mediterranean, Greece's position would be weakened. Similarly, if the Communists could gain enough say in Greek policy, then Turkey would be put in a weakened position.

With the containment strategy presented by Kennan fresh in their minds, the State Department decided the best strategy would be to funnel aid to both countries to calm down the long-running tensions between them. Defusing the situation would take away the potential leverage that could allow Communists to gain a foothold. These helped to show that the Truman Doctrine—an international intervention against the spread of communism (and Soviet influence)—had teeth.

The Truman Doctrine

The Truman Doctrine spelled out, in more formal and detailed terms, the perspectives and policies that were already demonstrated by the interventions in Turkey and Greece. The doctrine, which was publicly presented in March 1947, gave a pretty clear picture of how the United States saw the new world that was emerging. In this doctrine, each country was confronted with the choice between (a) the way of life that had representative government, free elections, and freedom of speech and religion and (b) the way of life that had government control over radio and the press, suppression of religion, and minority rule.

On the basis of this approach, the United States created a foreign policy focused on intervening in other countries to help make sure that Communism and Soviet power would be "contained."

The Marshall Plan

Truman made General George Marshall, a hero from World War II, his new secretary of state, and sent him to meet with other key leaders to push for agreements on creating a strong, stable Germany. Marshall met with his Soviet counterpart, foreign minister Vyacheslav Molotov, and others. The line of argument put forward was that to make Europe stable and to lift the economy overall, it was necessary to first make Germany stable and lift up its economy.

In order to help get Germany up and running again, one of the key things Marshall pushed for was a more complete picture of how many industrial plants and how many elements of the infrastructure had already been carted back to the Soviet Union. When Molotov still wouldn't go along with demands of the group (for more than five weeks of negotiating), he quit, and the meeting ended. Marshall had also met with Stalin to see if Stalin was any more favorable to a stable, profitable Germany. No luck.

On June 5, 1947, Marshall officially unveiled the plans for large-scale assistance to any countries in Europe who wanted to participate—even the Soviet Union and the

Eastern Bloc countries. Proposed in 1947 and implemented in 1948, this plan would pour a little over thirteen billion dollars into rebuilding European economies. The capitalist elements of the plan and the required cooperation among the participants were also seen as a way to help prevent the spread of Communism. It is probably no surprise that even though the Soviet Union was offered the chance to be included, they chose not to be.

Instead, in 1949, the Soviets created something called the Council for Mutual Economic Assistance. This council used trade agreements with the Soviet Union as a way to spark recovery among various Eastern European nations. It was also meant to be an alternative to the Marshall Plan.

Czechoslovakia and Yugoslavia

There was one country in Eastern Europe that had not been overtaken by Communism yet, and that was Czechoslovakia. They tried to get help from the Marshall Plan, but then changed direction in response to pressure from the Soviets. By February 1948, democracy in the country was gone. (So, too, was the foreign minister, Jan Masaryk, who died under strange circumstances.)

As the world looked to settle in for a long, tense Cold War, a couple of events made everyone stop and reflect on some assumptions that were being leaned on too heavily. The idea in the West that Communism is one shared set of views and beliefs somehow was changed by the rebellion in Yugoslavia and by its leader, Josip Broz (known as Marshal Tito).

The KGB

It has been said that the best way to understand the Cold War is not as a struggle between two political systems but between two intelligence agencies.

The Soviet KGB was created in 1954 and was an evolved model of organizations that had existed in different forms for many years. In Russian, it stands for *Komitet Gosudarstvennoy Bezopasnosti*. In English, it is known as the Committee for State Security. It was built to be used (and carefully controlled) by the senior officials in the Communist Party.

The KGB also became a useful tool allowing the Soviets to gather the various bits of research and technologies they needed to be able to upgrade their submarines and airplanes to stay competitive in the arms race. Perhaps nothing it achieved was quite as important as managing to steal important secrets that allowed it to accelerate Soviet research on atomic weapons.

As the Cold War unfolded, the KGB was becoming the counterbalance to America's CIA. There was one big difference, though: While the CIA was focused externally—just looking at foreign intelligence and counterintelligence—the KGB saw itself as dealing with dissidents and internal issues as well. They engaged in extensive monitoring of public and private expressions of opinion, not only within the Soviet Union itself, but also within the satellite states.

At its peak of close to five hundred thousand employees, the KGB was the largest such organization in the world—partially because it combined several kinds of organizations into one.

In discussing the Soviet control of the satellite states, it is worth noting the first of the challenges to Soviet **hegemony** from within the group of Communist countries. A Communist government had come into power in Yugoslavia without help from the Soviet Union. That is an important difference compared to most of the other Eastern European nations that were part of the Soviet flock. Because the leadership did not owe their rise to the Soviets and because there was very strong support for the Communist party, Tito was able to lead the federation from 1945 to 1980 (when he died) and resist having Stalin tell them what to do.

Was Stalin comfortable with that? Of course not! In 1948, he kicked Yugoslavia out of the Soviet bloc of nations. That allowed Tito to build better relations with other Eastern European nations and to establish connections with nonaligned countries. However, it would be a long time before another country was able to get away with exerting its independence and breaking away from Soviet domination.

In 1948, Tito and Stalin debated back and forth about who would determine Yugoslavia's direction and policies moving forward. In response to continuing tension over internal policy decisions, Tito cleaned house and swept out all the Yugoslav Communists who seemed to owe their loyalty to Moscow rather than Yugoslavia.

Stalin responded with threats and a big, showy series of public trials in other satellite nations to deter those countries from feeling inspired enough to follow Tito's lead. But the ploy was effective and, in the end, only Tito and Yugoslavia were able to continue making their decisions rather than march in step with Moscow.

The Soviet development of atomic weapons came in August 1949. This was the second major event, referred to above, that helped to kick away some fundamental assumptions that many people were relying on in their reasoning about the best path forward. With Stalin's passing in 1953, the new premier of the Soviet Union emerged—Nikita Krushchev. Many hoped there would be a kind of thaw in the Cold War at this point.

Au Revoir, France

While Germany joined NATO in 1955, France left just a few years after that. By 1958, there was more and more strain as de Gaulle continued to push back against the organization being driven mostly the United States. Some of his criticism also revolved around decisions and actions by NATO that he felt were a violation of France's right to make its own decisions—not least of which was de Gaulle's fear that other countries could decide whether or not France would be automatically dragged into another war.

In 1966, then, France made its formal exit from NATO and told NATO to take its soldiers and headquarters and leave France. While keeping diplomatic relations with NATO and committing to help in cases of unprovoked attack, it would not be until 2009 that France would rejoin to participate in command of NATO.

New Research and Technology

The launch of *Sputnik I* sparked a number of changes, including the creation of ARPA (Advanced Research Projects Agency, now DARPA, with the D standing for Defense). Initially created in 1958 to prioritize different space and technology projects, it grew to cover a wide range of technologies. The most significant development to come out of their research is the ARPANET.

ARPANET (shortened from ARPA and "Network") was the precursor to our modern internet. To understand what the innovative development was, we have to talk about packet switching. In the early days, if computers wanted to be able to communicate with each other, there had to be a connection path between those two machines. This worked like phone landlines did. As you can imagine, that would require switchboards on a massive scale and would make it too complicated to have large numbers of machines interacting with each other at the same time. However, by giving each machine a kind of mailing address, and breaking the information we wanted to send into small "packets" with the target address hooked onto it, then it could take any available route to get to the address.

This was an important development, especially for military and espionage applications, because the information could still be routed to the correct destination even if parts of the network had been neutralized or destroyed. Without it, we would not have grown from the handful of machines on the original network to our current level of over three billion around the world.

Continued Unrest in the Eastern Bloc

While there were relatively few real problems with NATO, the Warsaw Pact continued to show increasing signs that fault lines were growing. The part of the Warsaw Pact that allowed for Soviet troops to be stationed in the satellite nations sparked some nationalist anger in Poland and Hungary as they began wanting more say in their own fates. When a new Communist leader was swept into office following a series of strikes and riots, Krushchev met with him, and they were able to reach an agreement that included allowing Poland more freedom.

Any inspiration that revolution might have given to other countries about how far the swing toward freedom would be allowed to go was clamped down when Hungary attempted to leave the Warsaw Pact. From 1953 to 1955, Hungary was under the leadership of Imre Nagy, who was pushed out by the Soviets. In 1956, starting at the end of October, mishandling of a student demonstration by government forces resulted in the death of a student when they were fired upon by state police. The revolutionaries working against the Soviets appealed to Nagy, and he was put back in power. He promised things like freedom of speech in addition to withdrawing from the Warsaw Pact. Declaring Hungarian neutrality, he appealed to nations of the West for support, but they were too scared of sparking a war. Krushchev then crushed the revolution and, in 1958, finished the job by killing Nagy for treason.

Hungarian revolutionaries protest on and around a Soviet tank.

After challenges to Soviet control by Yugoslavia and Hungary, a third case would not show up until 1968 and would meet with failure much like Hungary's. After the Czechoslovak leadership had started allowing more freedom of expression and had reached out to connect with countries in the West, the Soviets invoked the Warsaw Pact, sent troops in, and crushed anyone they thought needed crushing to restore compliance.

Korea became a hot zone in the Cold War and was one of the first tests of the new United Nations.

CHAPTER FIVE

The Legacy of the NATO, the Warsaw Pact, and the Iron Curtain

The first peak in the on-and-off cycle of tension was from 1948–1953. That short span covered the Berlin Airlift, the creation of NATO, the success of the Soviet race to develop nuclear bombs, and, because it had been too long since anybody invaded anybody, the start of the Korean War.

There was a bit of easing in tension with the death of Stalin in 1953, which stretched to 1957. That being said, this period still saw West Germany admitted to NATO, and it saw the creation of the Warsaw Pact by the Soviets and the satellite states.

Warm and Cold Cycles

The next peak came between 1958 and 1962 when the fight to be on top really started to get serious. The United States and the Soviets started racing to see who could build the most intercontinental ballistic missiles. Then the Soviets tried to sneak some missiles past the United States and install them in Cuba (1962). This sparked what is known as the Cuban Missile Crisis because of the danger of having Soviet missiles parked in the United States's back yard. President John F. Kennedy, who was only in the second year of his term, told the Soviets, more or less, that if they didn't take their missiles back to the USSR, the United States would send its own missiles over there instead (except these missiles would land with a bang, instead of being carried off ships peacefully).

This incident helped both sides get a tiny bit of perspective on the bigger picture of not blowing up our only planet, and they moved to agree on signing the Nuclear Test Ban Treaty of 1963. The treaty banned the testing of nukes above ground. On the downside, though, the incident sparked the Soviets to start building up the stockpiles of non-nuclear weapons so they would not be smacked down so easily (and humiliated) next time.

For the next half-century, both superpowers managed to avoid having to smash their tanks and planes against each other. The only time they even used their tanks and fighters was to help keep allied countries from defecting to the other side (or to overthrow them if they *did* switch sides).

The Soviets stepped in to protect the Communist governments of East Germany in 1953, Hungary in 1956, Czechoslovakia in 1968, and Afghanistan in 1979.

The United States didn't roll in with guns out but helped overthrow governments. They helped topple the government of Guatemala in 1954. They supported the (failed) invasion of Cuba in 1961, and they invaded the Dominican Republic in 1965. They also engaged in a long (1964–1975) and unsuccessful effort to prevent North Vietnam, which was Communist, from taking over South Vietnam.

War in Korea

One of the more interesting outcomes of the formation of the NATO versus Warsaw Pact spheres of power was the outbreak of hostilities in Korea in 1950. This case is particularly interesting because while we normally think

A military convoy crossing the dividing line between the warring sections of Korea

of the Iron Curtain as a phenomenon of Eastern Europe, it is reflected here as well. At the close of World War II, the United States and the Soviet Union had drawn a dividing line across the Korean Peninsula at the 38th parallel. Territories on either side were established as independent countries with the north supported by the Soviet Union and the south by the United States and its allies, much like had happened in Germany.

When the leaders of the People's Democratic Republic of Korea (North Korea) attempted to reunify the country by force, the United Nations passed a resolution calling all member nations to help defend and protect the Republic of Korea (South Korea). Here, then, we see part of that careful and complicated relationship between the United Nations and the polarized world of the superpowers that was mentioned early on. The mess got messier when the forces backed by the United States pushed all the way into the north, and China decided to get involved as well. The whole story is too long and complicated to cover here, but basically everything ended up as a stalemate.

Toward A New World Order

The Soviet Union managed to launch the satellite *Sputnik I* into space on October 4, 1957. *Sputnik* was the first man-made satellite in orbit. In a way, this was like stretching the Iron Curtain all the way into space because now that was also an area of competition between the superpowers, and each was afraid to let the other gain an advantage. It is always easy to point fingers when you lose a race. The

The launch of the Soviet satellite *Sputnik I* into space sparked new fears and ignited the United States space program.

American public is as good at this as anyone, so it is no big surprise that there were a lot of fingers pointed at President Eisenhower and blaming him. Why had defense budgets been cut? Why hadn't he built up a space program? Be that as it may, it was clear that changes needed to be made in order to boost science and space research, so plans were put in motion that would lead to the creation of NASA in 1958.

This seems to be fairly characterized as the next peak in the Cold War, as both groups were now racing to build more missiles than each other, bigger and faster computers than each other, and farther-traveling, better-flying space vehicles than each other.

During the better part of the next two decades, though, the tug of war between the two superpowers got more complicated. The cozy friendship between the Soviets and China, for instance, hit a rough spot, and they started moving toward separate goals. Over that same period, both Japan and the countries of Western Europe had strong periods of economic growth, which closed some of the gap between them. That meant that it was now less clearly a seesaw only built for two.

The cycles were to go up and down another time or two before we finally put the worst of it behind us. In the 1970s, things eased somewhat, and the superpowers managed to get pen to paper on a couple agreements to limit how many missiles they really ought to build. Then in the 1980s, the tension ramped back up again as they competed to make friends and influence people in the third world.

Reagan, Gorbachev, and the End of the Warsaw Pact

There was a **resurgence** of tension in the 1980s, heightened in particular by the Soviet invasion of Afghanistan in 1979. The very next year, 1980, Ronald Reagan was elected president of the United States. Under Reagan, the United

Reagan (*left*) and Gorbachev (*right*) would oversee the apparent end of the Cold War.

States embarked on a period of increased military spending. The pace was too much for the already-strained economy of the Soviet Union to keep up with. By 1985, the new Soviet leader, Mikhail Gorbachev, admitted the economic problems in his country, and he worked to reach several arms-reduction agreements with the United States. He also launched important economic reforms.

Gorbachev came into office in 1985, and he was to prove an effective partner in working with Reagan to help move international politics past the Cold War era. Not only was he instrumental in initiating reforms within the Soviet Union itself, but he was also willing to let go of Eastern European countries when democratic governments rose to power.

In the summer of 1989, Gorbachev was obliged to recognize the independence of the various Eastern European countries. The end of support to prop up the Communist leaders in those countries cleared the way for them to be replaced by freely elected governments. As the Soviets relinquished their grip on the countries of central and Eastern Europe, along with the weapons reductions, the military role of the Warsaw Pact was at an end for all intents and purposes, and it was dissolved in 1991. Fittingly, this was one of the last steps on the path to formal dissolution of the USSR itself, which followed in December of that year.

With the success of revolutions in 1989, bringing democracy to the countries in Eastern Europe, the era of the Warsaw Pact seemed to have reached its end. At a final meeting in Prague, the leaders of the Warsaw Pact officially declared it "nonexistent" on the first of July in 1991.

Fall of the Berlin Wall

The outer fence along the border in Hungary was the first section to be taken down. It was the breaking of the giant concrete Berlin Wall, however, that really symbolized the turning point—just as its construction had. It went up in August 1961, and on November 9, 1989, the new democratic government of East Germany declared the border open (though it would take another month or so to fix everything to allow visa-free travel between the East and the West). That same evening people started chipping and breaking the wall. They were given the nickname of "Mauerspechte" or "wall woodpeckers."

NATO: Beyond the Cold War

Given the dissolution of the Warsaw Pact, and the centralized conglomerate of nations that had brought it to life, many questioned the need to maintain NATO as an organization—especially in its military role.

However, with membership being extended to Germany after its reunification (to replace the former West Germany in 1990), and the new roles and needs of nations like Poland, Hungary, and the Czech Republic, the need for NATO to remain and shift its focus to a more political role became apparent. In its new role, it would play a central part in stability and integration of Europe. By 1991, the new membership total rose to nineteen.

Then, in 2002, other countries joining the organization were former parts of USSR from the Baltic (Estonia,

Soldiers gather around remnants of the now-defunct Berlin Wall.

The Legacy of NATO, the Warsaw Pact, and the Iron Curtain

Lithuania, and Latvia), a former part of the independent Yugoslavia (Slovenia), and parts of the former Czechoslovakia (Romania, Slovakia, and Romania). Russia itself was even welcomed in as a limited partner.

This new role for NATO as an organization was to preserve and promote "cooperative security." The two primary goals included under this term were: building dialogue among former adversaries and managing conflicts (and potential conflicts) in the region around the European group—the Balkans, for instance.

To fulfill the goal of building dialogue and cooperation, NATO created something called the North Atlantic Cooperation Council in 1991. This council was designed to provide a channel for members to exchange their views and concerns—both political and military. One part of this was a program called the Partnership for Peace. It was launched in 1994 to set up joint military training operations between NATO and non-NATO nations.

The fulfillment of the organization's second new goal came to life in 1995 with NATO's first use of military force. It stepped into the war between Bosnia and Herzegovina. In particular, it flew air strikes against Bosnian Serbs who were positioned around Sarajevo, the capital.

It continued in its role there by stationing NATO peacekeeping units in the area after the signing of the Dayton Accords that brought a close to the open dispute.

Again, in 1999, NATO called in air strikes against Serbia in order to try and force the government, under its president Slobodan Milosevic, to go along with a diplomatic agreement, which was aimed at protecting the (largely

Piroska Bata: Life at the Border Wall

Meet Piroska Bata. Piroska is a professor of nursing who lives in Canada. She spent part of her childhood in Yugoslavia and has traveled back many times. Her profile will help us see a border wall from her perspective.

Do you remember the sights of driving up to the border walls and fences? What went through your imagination at those moments?

At the border was a long line. Always. My parents always gave us instructions to behave, and to stay quiet. Don't say anything. The guards on the sides and around us; some had German Shepherds, and all of them were wearing machine guns. Guns were everywhere. It was very intimidating.

My most scary memory was when my father became ill while we were waiting. He needed to throw up and ran into the bushes to do so. Suddenly I saw soldiers with guns and dogs chasing him, but when they saw him throwing up they just stood guard.

What stands out to you as the biggest change before and after when the walls and fences came down? How do you see your experience now as an adult?

Our culture stayed strong even while it was hidden. I know that we had this record that my dad purchased in Germany, and we were not allowed to play it. It was contraband or illegal. Now when I go back to visit, there are colorful traditional arts and the joyful folk music mixed in with things.

Muslim-Albanian) population of Kosovo. To help maintain the conditions of the settlement, a peacekeeping force called the Kosovo Force was deployed.

In the aftermath of the Kosovo intervention, there was some debate among countries in the European Union (EU). This was about possibly building up a specific EU force for intervention in crises. It was not clear whether this would strengthen NATO or weaken it. As the world moved into the second decade of the twenty-first century, it seemed that the EU would not really be able to develop a crisis-intervention force on the scale of the one that could be deployed by NATO.

Into the Twenty-First Century

In the late 1990s and early 2000s, the United States worked to bring more countries under the NATO umbrella. Among the countries reached out to were former Soviet allies. This was seen as helping in two ways: firstly, it helped create alliances that would secure the newly gained freedom from under the thumb of Russia. Secondly, it would help plug them into regional policy and economic discussions.

Even though there is still active debate about the future direction of the organization, the organization now includes such countries as Hungary and the Czech Republic. What is more, Russia has formed a cooperative relationship with NATO to help work together on shared concerns like arms control, terrorism, and so on.

It seems that several interesting developments will begin to overlap as trends continue. For instance, what sense will it make to continue spending time and resources to sustain both the United Nations and NATO? Similarly, how much of a role will geography even continue to play in trying to draw any meaningful dividing lines (a lesson that military forces are already being forced to learn)?

Yet, even as this book goes to press, there is renewed tension between the United States and Russia over election interference, and there are disagreements between them over the correct intervention strategy in Syria. Perhaps we should be careful not to think there will be no more Cold Wars with the same naiveté as the people in the aftermath of World War I imagined there would be no more wars period. This is more than a wide philosophical speculation, however, because how we understand these tensions will influence future diplomatic relations.

Chronology

1945   The conferences at Yalta and Potsdam take place. World War II ends in Europe. Atlee replaces Churchill.

1946   Kennan's "Long Telegram" is sent. Churchill makes his "Iron Curtain" speech.

1947   The Truman Doctrine is established. The CIA is formed.

1948   The Berlin Airlift begins.

1949   NATO is founded. The Federal Republic of Germany (West Germany) and the German Democratic Republic (East Germany) are formed.

1950   The United States enters the Korean War.

1953   Death of Stalin.

1954   The KGB is formed.

1955   West Germany enters NATO. The Warsaw Pact founded.

1957   *Sputnik I* is launched.

1961   The Berlin Wall goes up.

Glossary

advocacy Supporting, recommending, or arguing in favor of something; working to get support or agreement for something.

ambassador A person who acts as the representative for someone, for an organization, or for a specific activity.

Bolshevik One of the groups with differing ideas of communism; the Bolsheviks are the group that gained control in the Russian revolution.

Communism A theory that the best way to organize a society is for the government to own everything, each person works to contribute as much as they can, and the government provides them with whatever they need.

diplomat Sometimes another word for ambassador; usually used when the person or group being represented is a government.

embargo When a government stops (or requires others to stop) some activity, like buying any goods or supplies from Country X or Country Y.

espionage Spying or using spies; especially when it is to gain military or political secrets from another country.

exile When your home country takes away your legal permission to live there; being forced to leave.

grandeur The quality of being large and beautiful or special; being impressive; could be literal or metaphorical.

hegemony When one group or person has way more control and influence over something than the others.

ideological Having to do with a group of ideas or beliefs.

industrialization Changing an economy that is based mostly on farming to one based mostly on manufacturing.

intelligence One meaning of this term is information that has political or military value.

monarchy A system of government that has a royal person, like a king or queen, as the highest decision-maker.

renaissance A period of rebirth or revival; usually used to refer to a sudden increase in improvements in fields of study ranging from art and music to science.

reparations Money a country is required to pay to help make up for the damages and expenses it caused in a way.

resurgence When something has increased growth or activity level again after a low period.

schism When a group or organization splits into two or more separate groups; usually used when the division happens over differences in ideology.

Slavic Refers to countries and cultures that speak a related group of languages; examples include Russian, Polish, Czech, Serbian, etc. Interesting to note that Hungarian is not a Slavic language, though Hungary is often included in references to Slavic countries.

Soviet From the Russian word for council; Soviet Union refers to being composed of many soviets, or councils of workers (including peasants and soldiers); often used to refer to citizens or the government leaders of the Soviet Union.

Bibliography

Applebaum, Anne. *Iron Curtain: The Crushing of Eastern Europe 1944–1956*. New York: Random House, 2012.

Bentley, Jerry H., and Herb F. Ziegler. *Traditions and Encounters: A Global Perspective On the Past*. New York: McGraw-Hill, 2006.

Best, Geoffrey. *Churchill: A Study in Greatness*. New York: Hambledon and London, 2001.

Holloway, David, and Jane M. O. Sharp, eds. *The Warsaw Pact: Alliance in Transition?* Ithaca: Cornell University Press, 1984.

Huston, James A. *One for All: NATO Strategy and Logistics through the Formative Period (1949-1969)*. Newark: University of Delaware Press, 1984.

Ismay, Lord. *NATO: The First Five Years 1949-1954*. Netherlands: Bosch-Utrecht, 1954.

Jones, Christopher D. *Soviet Influence in Eastern Europe: Political Autonomy and the Warsaw Pact*. New York: Praeger Publishers, 1981.

Kennan, George F. (X). "The Sources of Soviet Conduct." *Foreign Affairs* 25, no. 4 (July 1947): 568.

King, Margaret L. *Western Civilization: A Social and Cultural History.* Upper Saddle River, NJ: Prentice-Hall, 2006.

Luns, Joseph. Preface to *NATO The North Atlantic Treaty Organization: Facts & Figures.* Brussels: NATO Information Service, 1976.

Miller, David. *The Cold War: A Military History.* New York: St. Martin's Press, 1998.

Pittaway, Mark. Brief Histories: Eastern Europe 1939–2000. New York: Oxford University Press, 2004.

Ramet, Sabrina P., ed. *Eastern Europe: Politics, Culture, and Society Since 1939.* Bloomington: Indiana University Press, 1998.

"Stalin's Reply to Churchill" (interview with *Pravda*), *New York Times,* March 1946, 4. Accessed October 21, 2016. http://sourcebooks.fordham.edu/halsall/mod/1946stalin.html.

Stearns, Peter, Michael Adas, Stuart Schwartz, and Marc Gilbert. *World Civilizations: The Global Experience,* vol.2. New York: Addison-Wesley Educational Publishers (Pearson), 2001.

Further Information

Books

Brands, H.W. *The Devil We Knew: Americans and the Cold War.* New York: Oxford University Press, 1993.

Gaddis, John Lewis. *The United States and the Origins of the Cold War, 1941-47.* New York: Columbia University Press, 1972.

Kennedy-Pipe, Caroline. *Russia and the World, 1917–1991.* London: Arnold, 1998.

Willis, Jim. *Daily Life Behind the Iron Curtain.* Santa Barbara, CA: Greenwood Press, 2013.

Websites

The Cold War Museum

http://www.coldwar.org/

Presented by the Cold War Museum located in Vint Hill, Virginia, this website has brief articles on different topics, photo galleries, a fun Cold War trivia game, and several online exhibits.

The Library of Congress

http://www.ibiblio.org/expo/soviet.exhibit/soviet.archive.html

This website presents information about the Soviet Union exhibit at the Library of Congress, including some of the darker aspects and what was happening behind the scenes.

PBS: Public Broadcasting Service

http://www.pbs.org/weta/faceofrussia/intro.html

This website allows you to go in and learn more about the rich tradition and achievements of Russian culture, which is a valuable way to gain an understanding of the Cold War from both sides.

Videos

The Beginning of the Cold War

http://ny.pbslearningmedia.org/resource/pres10.socst.ush.now.coldwar/the-beginning-of-the-cold-war/

This video from PBS presents an examination of the ways Truman sought (from the beginning) to put strategies and resources in place to prevent the spread of Communism and the USSR.

The Cold War over CNN's Cold War

https://cosmolearning.org/documentaries/cold-war-by-cnn-perspectives/1/

This video is part of the CNN Perspectives series, which includes a discussion of the Iron Curtain and the Marshall Plan.

Index

Page numbers in **boldface** are illustrations. Entries in **boldface** are glossary terms.

advocacy, 59
ambassador, 68

Berlin Airlift, 32, **34–35**, 87, 96–97
Bolshevik, 52

China, 12, 32, 74, 90, 92
Churchill, Winston, 9, 18, **19**, 20, 23, 28–30, 56–57, 62, 64, 71
CIA (Central Intelligence Agency), 7, 17, 66–68, 79
Communism, 17, 25, 31, 41, 52, 57, 76–78
Communist Party, 22, 25, 48, 52, 79–80

de Gaulle, Charles, 20, 57, **58**, 59, 81
diplomat, 17, 53, 66–67, 81

embargo, 33

espionage, 68, 82
exile, 25, 68

Germany, 6, 15–16, 18, 20, 22, 33, 36–37, 39, 43, 45, 52, 57, 60, 64–65, 71, 77, 81, 87–88, 90, 95, 99
Gorbachev, Mikhail, 71, **93**, 94
grandeur, 59
Great Britain, 6, 9, 11–13, 15–16, 18, 21, 23, 28–30, 33, 36, 42–43, 74
Greece, 6, 9, 43, 74, 76

hegemony, 80
Hitler, Adolf, 15–16, 18, 21, 48, 57

ideological, 27, 37, 40, 59
industrialization, 16, 52, 54
intelligence, 9, 17, 51, 67, 79
Iron Curtain, **24**, **26**, 27–29, 31–32, 36, **37**, 39–40, 90

Japan, 12–13, 17, 21, 23, 62, 92

Kennan, George, 53, 65–68, **65**, 76
Korea, 12, 63, **86**, 89–90, **89**

Lenin, Vladimir, **4**, 13, **19**, 52, 68, 70

Marshall Plan, 64, 77–78
Marshall, George, **63**, 63–64, 67, 77
Molotov-Ribbentrop Pact, 16, 71
Molotov, Vyacheslav, 32–33, 68, **69**, 70–71, 77
monarchy, 76

NATO, 9, 27–28, 32, 41, **41**, 45–46, 59, 64, 67, 81, 87, 89, 95, 98, 100–101

OSS (Office of Strategic Services), 17, 66

Poland, 6, 14–16, 20, 23, 29, 31, 46, 64, 83, 95
Potsdam Conference, 18, 20, 23
prison camps, 25, 48, 54

Reagan, President Ronald, 93–94, **93**
renaissance, 12
reparations, 20
resurgence, 93
Roosevelt, President Franklin D., 18, **19**, 20, 23, **50**, 54–55, 62, 67

schism, 25
Siberia, 54, 68
Slavic, 12, 18, 36
Soviet Union, 67–68, 71, 73–75, 77–81, 90, 94
Soviet, 5–7, 13, 15–18, 21–24, 29–33, 36–37, 41–42, 45–46, 48, 53–54, 60, 62, 66–71, 74–79, 80–81, 83, 85, 87–88, 92–94, 100
Sputnik I, 82, 90, **91**
Stalin, Joseph, 6, 9, 13, 15–16, 18, 20–21, 23, 25, 29–31, 36, **47**, 48, 51–54, 62, 68, 70–71, 77, 80–81, 87,

Truman, President Harry S., 9, 23, 28, 32, 60, **61**, 62–64, 68, 75, 77

United Nations, 9, 20, 24, 60, 71, 74, 90, 101
USSR (Union of Soviet Socialist Republics), 6, 8–9, 13, 16, 25, 75, 88, 94–95

Warsaw Pact, 9, 27–28, 45–46, 71, 83, 85, 87, 89, 94–95
World War II, 5–7, 11–12, 15, 17–18, 29, 36, 52, 54–55, 57, 59, 64, 74, 77, 90

Yalta Conference, 18, **19**, 20

Index   111

About the Author

Erik Richardson is an award-winning teacher from Milwaukee, where he teaches a variety of college courses, including work for the graduate department of political science at Marquette University. He holds a position in city government, and his writing and consulting projects include work with the Foreign Military Studies Office of the United States Army, with the United States Department of the Navy, and with the Naval Postgraduate School.